One Cow and Counting

By Page McBrier   Illustrated By Regan Dunnick

PASS ON
THE GIFT

HEIFER®
INTERNATIONAL   www.heifer.org

One Cow and Counting

Page McBrier, Author
Regan Dunnick, Illustrator
Produced by Heifer Project International

This book was printed with soy inks on paper
containing at least 10% Post-Consumer Waste.

Printed in the USA
Signature Book Printing, www.sbpbooks.com

One big idea can have far-reaching benefits. So it is with the work of Heifer International, a non-profit
organization whose mission is to end hunger and poverty and care for the Earth. Heifer approaches this mission by
supplying intensive, Earth-friendly agricultural training and farm animals to poor families around the world. Since
1944, Heifer International has distributed animals ranging from earthworms to water buffalo, goats to geese and
cows to camels. As of this writing, Heifer International has helped more than 20 million families achieve lives of
independence, dignity and self-reliance. The impact of each gift is multiplied as families agree to Pass on the Gift
of their livestock's offspring to another family in need, widening the circle of hope.

Families who receive animals from Heifer learn how to grow plants that can Benefit the Soil and be used as ANIMAL FODDER.

more grass table five.

Many Heifer Goats and COWS are ZERO-GRAZED. The food is brought to their Pens.

9

Three Cows and Counting...

More calves meant more families had fertilizer for their crops and Gardens

One Cow Produces 50-100 pounds of MANURE a day.

corn

manure GOES here

Manure improves soil quality.

Other *animals* in the Manure Hall of Fame include chickens, goats, water buffalo, worms, pigs, horses and camels.

Manure hall of Fame

I Want to Thank...

clap clap

sheepbox Derby

some socks would be nice for those chilly mornings...

Hmm...

The Gift idea soon included other animals like SHEEP, whose wool makes warm clothing.

One year's growth of FLEECE is about 8 lbs of WOOL.

sheep tallow (fat) is used to make candles and soap.

Go sheep RACER, Go!

GOO GOO...

Lamb Meat is high in vitamins and minerals.

Young sheep are called Lambs.

13

They are also valued for MilK, Meat,

Hides,

MANURE and Horns

Mine too.

Not this Horn →

Chickens are omnivores; they eat both plants and animals.

Alektorophobia IS a fear of CHICKENS.

Ham or Green beans Hmmm... I'll have both!

BOO

Ekk!

Baconophobia IS a made up word for fear of BACON.

GRRR!

EKKK!

Worldwide, women earn 70-90% of what **men earn.**

NO ODOR!

BioGas Comes from methane, which is released from animal manure. DOES. NOT. SMELL!

Yes!

Biogas can save women 2-3 hours per day gathering firewood.

Women had money to buy medicine and medical supplies...

Almost half the world—
2.4 Billion people—live
on less than $2.00
a day.

The leading cause of
death among children
is pneumonia, an easily
treatable disease.

Insecticide-
treated bed
nets help
prevent
malaria,
a leading
cause of
DEATH.

BZzzz

UGH

BZ

Yes!

In INDIA, Heifer Women's Self-help groups used their community funds to build rooftop rainwater harvesting systems.

In China: women participated in group savings accounts to have money for emergencies or business start-ups.

Until recently, many more boys in the developing world attended school than girls.

Today the gap between boys and girls is only 3%.

Together, families and neighbors could work toward common goals.

In Cambodia, self-help groups give new project families a huge head start when they pass on cattle, swine, chickens, vegetables, seeds, fruit tree seedlings, animal feed *and* rice.